Contents

What is aerial photography?

Aerial photographs are those photographs that are taken from the air. They can be taken from ground-based sources, such as high-rise buildings and bridges, or aeroplanes, helicopters and even kites.

DIFFERENT TYPES OF PHOTOGRAPHY

The scale of aerial photographs can vary from very focused, small-scale images such as a particular house, to much larger expanses such as a whole town or city. The photographs can either be taken vertically above the point of interest or at an angle. Those that are taken from directly above are called vertical aerial photographs. Those that are taken from an angle are known as oblique.

Aerial photography allows us to view things from a different angle.

USING AERIAL PHOTOGRAPHY

Aerial photography was first used during World War I (1914–18) and since then it has become a widely used tool by a number of different groups of people.

Often a mixture of both oblique and vertical aerial photography is used. For example, construction companies use aerial photography to survey a site before it is developed. They can also take photographs to check and manage the progress of the building site. Environmentalists use aerial photography to assess the impact on the environment that things such as buildings cause. Cartographers – the people who draw and design maps – use aerial photography as the basis for their maps. Other uses of aerial photography include flood mapping and coastal surveys, both of which are vital in areas where flooding is a possibility.

FINDING SPECIFIC FEATURES

This book looks at many of the features geographers study and how to identify them on aerial photographs. Identifying features on an aerial photograph takes a bit of practice but once you know what to look for, aerial photographs can help you understand how the features were formed.

Have a look at the oblique aerial photograph on page 6. Can you identify what the numbers 1–6 are? Give reasons for your answers.

HELPING HAND

Throughout this book, this helping hand will give you useful tips and hints.

A photographer takes an aerial photograph. He is using a map as a reference to make sure that his photograph is accurate and of the right area.

Now, look at the list of answers below and see how you well you did:

- 1: main road
- 2: sports field
- 3: houses
- 4: side road and parked cars
- 5: public gardens
- 6: main intersection (roundabout)

KEY SKILLS

Throughout this book, you will learn different skills. Each different skill is represented by one of the following icons:

Completing a practical activity

Analysing information

Working with graphs, maps,diagrams and photographs

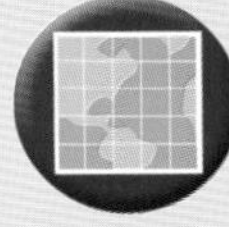

Looking at global issues.

Researching information

Observing

Settlements

A settlement is a place where people live. Some settlements go back hundreds or even thousands of years, while others are relatively recent. Settlements vary in size from tiny hamlets, with just a few houses, to huge cities with thousands of buildings.

WHY ARE VILLAGES FORMED?

Many considerations have to be taken into account before a settlement can be built – for example, the availability of good drinking water, a place where a river can be crossed safely, whether the river is likely to flood, whether there is a good view of the surrounding countryside, or whether the land is suitable for building on.

Make a note of why you think the settlement in which you live developed.

VILLAGES CHANGE

Villages are constantly changing. Some villages died out in the past because of changes in farming methods or disease that killed either people or farm animals. Today, many villages are expanding because people working in busy cities want to live in peaceful villages and commute to work each day. Can you think of any problems this might cause?

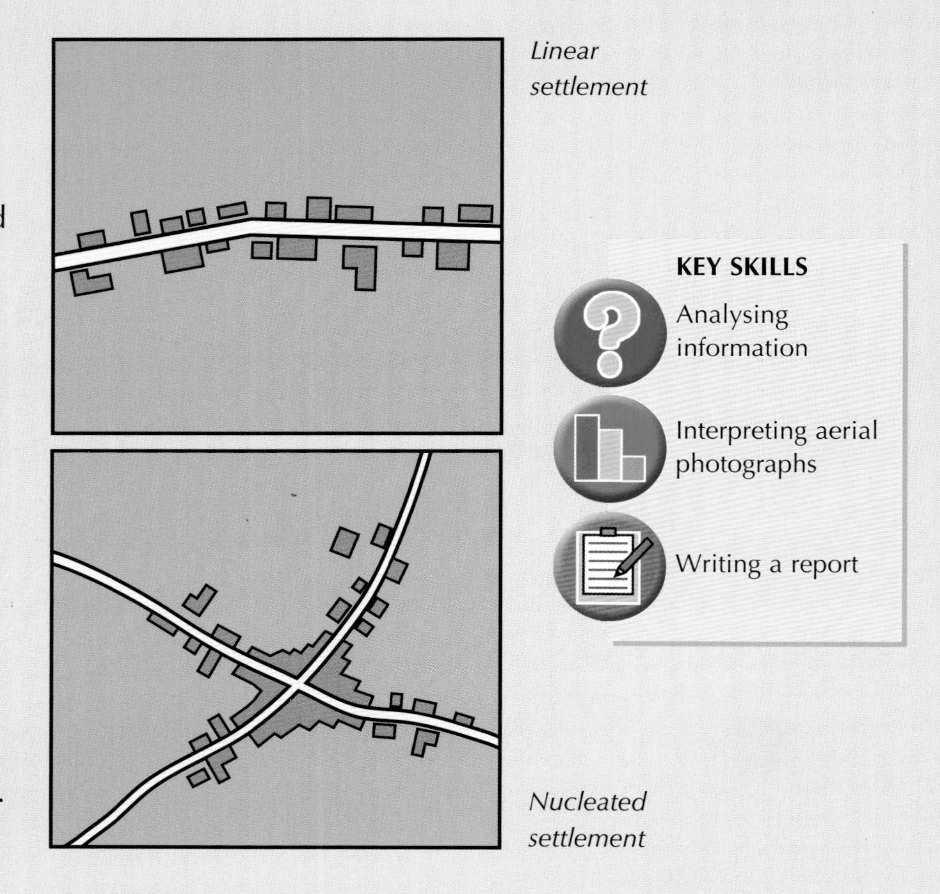

Linear settlement

Nucleated settlement

THE SHAPE OF THE SETTLEMENT

Most settlements divide into either linear or nucleated settlements. This describes the shape of the settlement. A linear settlement is one where the buildings have been built along a valley bottom, or on either side of a road, creating a long, narrow-shaped settlement. A nucleated settlement is one where all the buildings are grouped together, originally perhaps around a church or road junction. From the air, it is often clear which type of settlement you are looking at.

GEOGRAPHY SKILLS

Looking at AERIAL PHOTOGRAPHS

Helen Belmont

W

FRANKLIN WATTS

First published in 2006 by
Franklin Watts
338 Euston Road
London NW1 3BH

Franklin Watts Australia
Hachette Children's Books
Level 17/207 Kent Street
Sydney NSW 2000

Editor: Jennifer Schofield
Consultant: Steve Watts
(FRGS, Principal Lecturer University of Sunderland)
Art director: Jonathan Hair
Design: Mo Choy
Artwork: Ian Thompson
Picture researcher: Kathy Lockley

AA World Travel Library 31, Aerofilms/Scie … Lonely Planet Images 34, Adrian Arbib/
Alamy Images 13, Yann Arthus-Bertrand/CORE … es Ltd/Alamy Images 18, Michael Diggen/
Alamy Images 40, Mark Edwards/Still Pictures 30, Ow … national and © Infoterra 2006 3b, 9, 38, COVER,
Robert Harding Picture Library 20, Dennis John … S. Kirk/Still Pictures 23, Litografica Artistica
Cartografica 28, Vincent Lowe/Alamy Im … M-Sat Ltd/Science Photo Library 32,
Diego Lezama Orezzoli/CORBIS 36, c.Hubert Raguet/Eurelios/Science Photo Library 7, Dr Morley Read/Schience Photo
Library 15, Charles E. Rotkin/CORBIS 37, Dennis Scott /CORBIS 3T, 4–5, COVER, Jayanta Shaw/Reuters/CORBIS 17, Douglas Stone/
CORBIS 35, Jim Sugar/CORBIS 43, Jack Sullivan/Alamy Images 21, transit/Still Pictures 39, Tom Treick/CORBIS 26, University
of Cambridge, Collection of Air Photos/Science Photo Library 22, David Wall/Alamy Images 6, Jim Wark/Still Pictures 42,
Gerry Weare/Lonely Planet Images 12

A CIP catalogue record for this book
is available from the British Library.

ISBN 10: 0 7496 6781 8
ISBN 13: 978 0 7496 6781 8
Dewey Classification: 526.982

Printed in China

Franklin Watts is an imprint of Hachette Children's Books.

LOOKING AT PICTURES

Look at this vertical aerial photograph of a village. What type of settlement is it – linear or nucleated? Can you find the main crossing? Why do you think the settlement was formed? Look at the land around the village. What is it used for? Write a brief report on this village and why you think it developed.

HELPING HAND
The focus of a nucleated village could be the road junction, river, green or church.

This village developed around a road crossing. How could you use this vertical aerial photograph to draw a map?

Towns and cities

Towns are bigger settlements than villages, housing a much larger number of people and having more facilities, such as shops. For a settlement to have grown from a village into a town there has to be a reason.

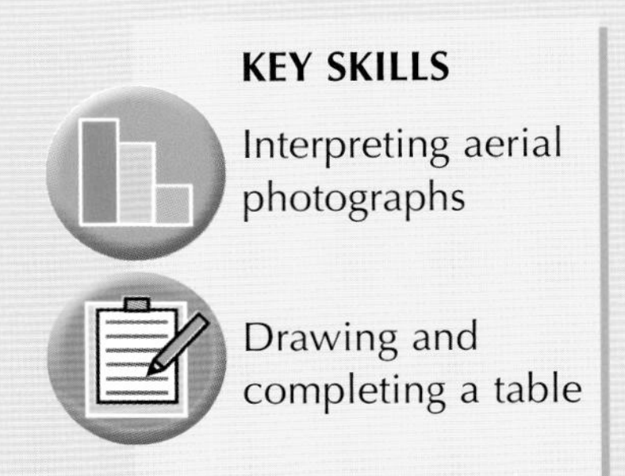

REASONS FOR GROWTH

Settlements can grow for a number of reasons: perhaps a new factory has been built and houses are needed for the workers, or the settlement has a deep river running through it, giving rise to a port for the transport of goods. It may be positioned close to a big town or city but provides a better place to live for city workers than being in the city itself. Often small towns develop into larger towns and eventually cities.

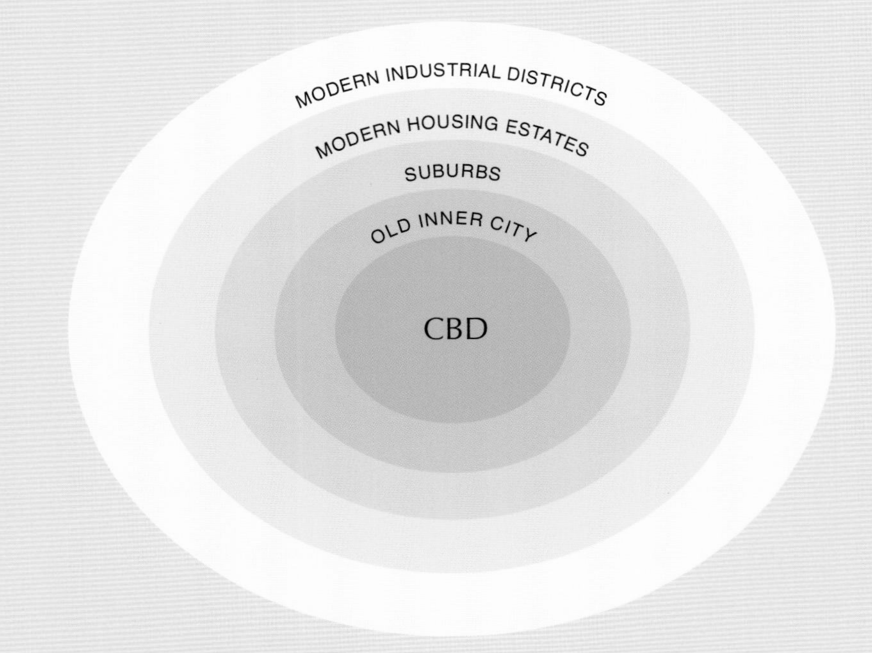

This is a diagram of the different zones of the city. The central business district is usually in the centre of the city with the industrial factories and buildings on the outskirts.

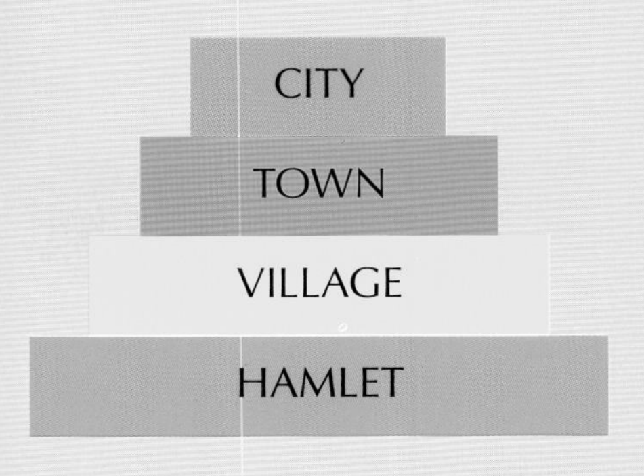

Settlements can be organised in order of their importance to form a settlement hierarchy. There are fewer cities so they are at the top of the pyramid.

THE DIFFERENT ZONES

Many old cities can be roughly divided into the following zones, or areas. The oldest part of the city is usually in the centre. The centre is called the central business district (CBD). This is where you will find most of the shops and other facilities. Next to the CBD is the old inner city, made up of small houses and gardens and usually an established industrial area, such as a factory or factory site. Further out are the suburbs, larger houses with gardens and small convenience shops. Around the edges of the city there are large modern housing estates. On the edge of cities you will find new industrial districts, which have easy access to the road network.

ON PHOTOGRAPHS

Using the diagrams on page 10 as your starting point, look at this aerial photograph. Can you identify the different zones? Can you see the ring road that goes around the city? Can you see houses joined together in a grid formation? These will be the terraced houses of the old inner city. Where are the houses with larger gardens and more space?

ZONE	POSITIVE POINTS	NEGATIVE POINTS
CBD	Close to many facilities	Busy and overcrowded Lack of greenery

Think about each zone and what you think it would be like to live there. Draw up a table like the one above, but listing all the zones.

Complete the table, listing the positive and negative points about each zone – the first one has been done to help you.

The circle you can see in this vertical aerial photograph is an inner ring road that goes around the CBD.

Village and city living

India is on the continent of Asia. It is a less economically developed country (LEDC), while countries such as the UK and the USA are more economically developed countries (MEDC). There are more than one billion people living in India. Where do they all live?

KEY SKILLS

Interpreting aerial photographs

Analysing information

Comparing two settlements

Using the Internet and reference books for research

The village of Dwali is in a range of mountains in India. What would this village look like on a map? *(See an example on page 28.)*

VILLAGE LIFE

Most people in India live in villages. Often these people are subsistence farmers. This means that they grow enough food to feed their families, and do not have much money spare for anything else. If the crops fail, the family will have no food.

Look at the oblique aerial photograph of the Indian village Dwali on page 12. Look at how the houses have been built. What sort of materials have been used? What do you think it would be like to live in this village? What sort of transport links do you think the settlement has?

MIGRANTS TO THE CITY

Many of the young adults who live in villages such as Dwali have had enough of living in difficult conditions, so they move to towns or cities, such as Kolkota and Mumbai, to find work. This is called migration and the people who move are called migrants. So many young people are migrating to the cities of India that there are not enough houses for them. The other problem is that often the migrants arrive without sufficient money to find a place to stay. As a result, they build homes next to their place of work out of any materials that they can find.

Use the Internet or go to the library to find information on living in Dwali and living in Mumbai. It may be useful to look for photographs of the two settlements. Draw a table like the one below to compare the two settlements.

	Dwali	Mumbai
Schools		
Leisure		
Working hours		
Job opportunities		
Healthcare		
Homes		
Running water		
Electricity		

HELPING HAND

Start your research by looking at www.mumbainet.com and www.india9.com for information about the two settlements.

This is Mumbai, one of India's most populated cities with over 13 million people living there. Some of Mumbai's inhabitants live in high-rise buildings but others have no choice but to build their own houses on the outskirts of the city in informal settlements.

A journey down a river

Rivers are a vital link in the water cycle, collecting some of the water that has fallen as rain or snow from the high ground and bringing it down to flow back into the sea. Some rivers stretch for hundreds of kilometres from their beginning, called the source, to their end, called the mouth.

V-SHAPED VALLEYS

At the source of the river, the river is shallow and quite narrow. As the river tries to move to the mouth, it wears away the rocks that it flows over, causing a steep V-shaped valley. Often you can see large boulders on the stream bed.

THE MEANDERING RIVER

As the river moves away from the source, it changes shape. This is because it flows over flatter ground. It creates a channel that is meandering, which means it has bends.

Look at the photograph on page 15. On the outside bend, the water is flowing fastest, cutting away at the banks of the river. The water flowing on the inside edge will have virtually stopped moving and so sediment is deposited. The sediment deposits are the pale patches on the inside bends.

A river basin

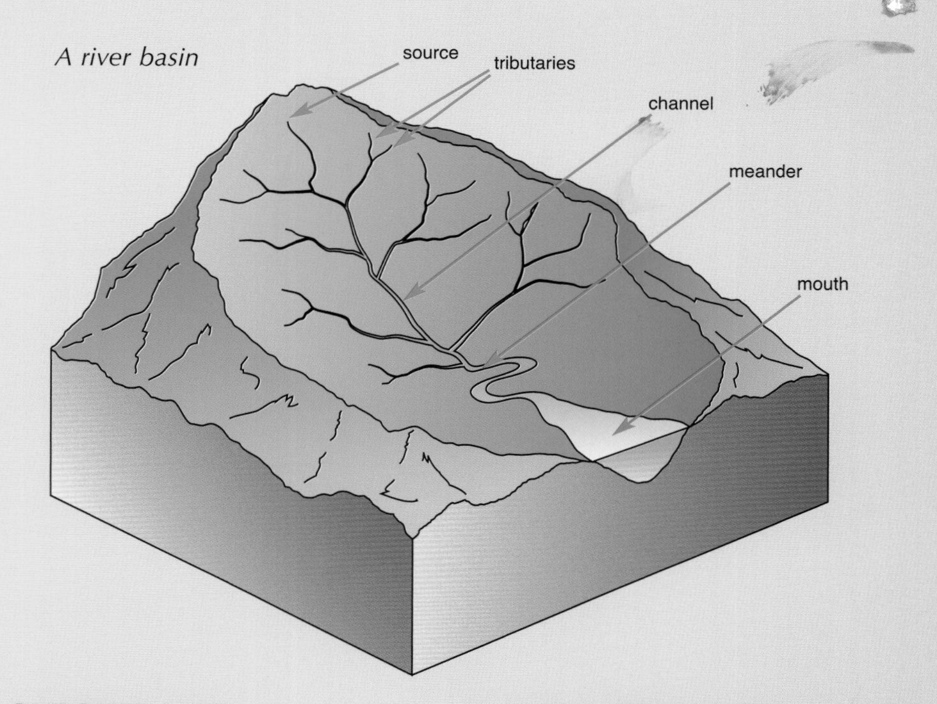

OXBOW LAKES

During a flood, the river often breaks through the neck of land between two meanders. The river then takes the shortest route, eroding a new channel and depositing material, cutting off the old meander. The old meander is called an oxbow lake. The horse-shoe shape in the bottom left corner of the photograph is an oxbow lake.

THE RIVER MOUTH

The river mouth is wide where it meets the sea, as all the channels of the river collect here. It is also deep, allowing ships to transport goods.

Using the photograph opposite, draw a sketch of the meanders and label the different features. How might the river change in the future?

The deposited sediment, or point bars, can be seen clearly on this photograph of the River Curaray in Ecuador.

The floodplain

At the mouth of a river, the flat area of land that falls either side of the river is called the floodplain. The floodplain holds the water when the river floods. A river floods when there is so much water that it overflows onto the surrounding land, before it can reach the sea.

KEY SKILLS

Interpreting aerial photography

Interpreting information

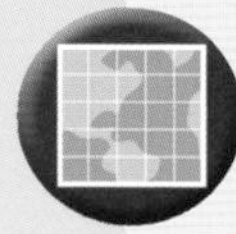
Finding out about Bangladesh

Writing an article

Using the Internet and atlases to do research

How might a picture like this help people to plan flood defences?

LOOKING AT BANGLADESH

In Bangladesh, most of the land is only one metre above sea level and the country is criss-crossed with rivers that start in the Himalaya mountains. Bangladeshis expect floods every year, especially after the monsoon rains between June and October when vast amounts of rain can fall every day.

USING THE FLOODPLAIN

Farmers grow crops on the floodplain. They depend on floods to keep the land fertile because after the floodwater has mostly drained away, it leaves rich mud, called silt, on the land. Farmers then use this mud to fertilise their farmland.

DESTRUCTION

Although floods can help the farmers, they can also cause massive destruction. Look at the photograph above. Can you see the traces of farmland that have been flooded? What about people's homes? Can you see any clues as to whether they were damaged by the flood?

These roads in Bangladesh have been flooded. This means that the people who live here are cut off from most outside help.

Severe flooding in Bangladesh, which has already killed nearly 400 people and caused hundreds of millions of dollars worth of damage, is set to get worse. Officials say high tides are affecting rivers in the centre of the country, and floodwaters, which are already waist-deep in places, will continue to rise. Over 700,000 hectares of farmland have been flooded. The water has been on the ground for more than a month in many areas. Many people have run out of money to buy food and cannot work because much of the landscape is submerged.

Imagine you are a reporter interviewing people after this flood. Use the news article and the aerial photograph above as your starting point to write a newspaper story about what the flood has meant to different people and what needs to be done in the future. Did they lose family members? Should there be more warning about the floods? Use an atlas and the Internet to do your research.

HELPING HAND

Go to: http://earthobservatory.nasa.gov/NaturalHazards and click on "Floods" then "Floods in Bangladesh".

Waterfalls and river deltas

As a river travels from its source to its mouth, it wears away or erodes the land it is passing through. The erosion can create waterfalls, or when the particles of earth and rock are carried down the river to its mouth, they can create landforms such as deltas.

KEY SKILLS

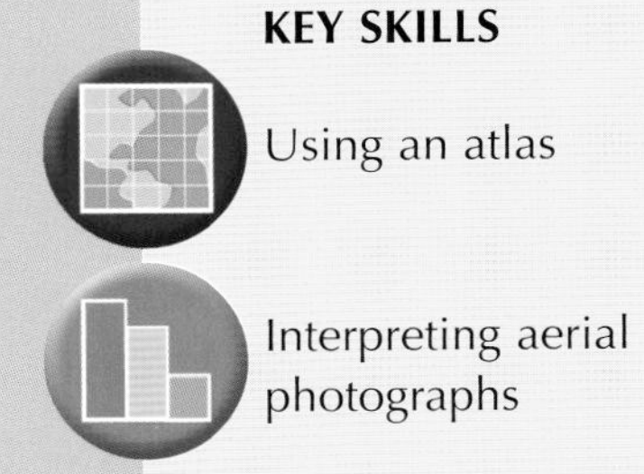

Using an atlas

Interpreting aerial photographs

Drawing accurate sketches

WATERFALL FACTS

Look at the oblique aerial photograph of the Niagara Falls. This waterfall lies on the border between the United States of America and Canada. Use an atlas to find Niagara Falls and to see how high it is.

Most waterfalls form when a river is flowing over a mixture of hard and soft rock. The river flows over the hard rock but erodes the soft rock, gradually wearing it away to create a drop in the water level. Can you see evidence of erosion in the photograph?

Niagara Falls is located on the Niagara river, which is over 50km long. The Falls may not be the highest in the world, but it is one of the most powerful.

GORGES

Below the waterfall, where the rock is eroded, a steep-sided valley, called a gorge, forms. Can you see the valley walls in this photograph? Over time, the gorge will become longer and longer, as more of the hard rock wears away.

RIVER DELTAS

As the river approaches the sea, at the river mouth, it starts to slow down. If it goes really slowly, the sediment, made up of small bits of rocks and earth washed down the length of the river, drop to the bottom of the river bed, where it can build up into islands of mud to form a delta.

THE NILE RIVER DELTA

Egypt's Nile river has a vast delta. Use an atlas to locate it. Now, look at the satellite aerial photograph below. Can you see the triangle shape delta? Over time, the river has split into different channels that flow around mud islands. These channels are called distributaries. Use the aerial photograph below as a guide to complete a sketch of the delta. Label the distributaries, floodplain and desert. From the photograph, you can see that the delta is very green. Most of Egypt's farming is done here.

Geographers use aerial photography to monitor the Nile delta as its supply of sediment has been reduced by a huge dam upstream. The farmers now need to use artificial fertilisers on their crops where before, the sediment did the job for them.

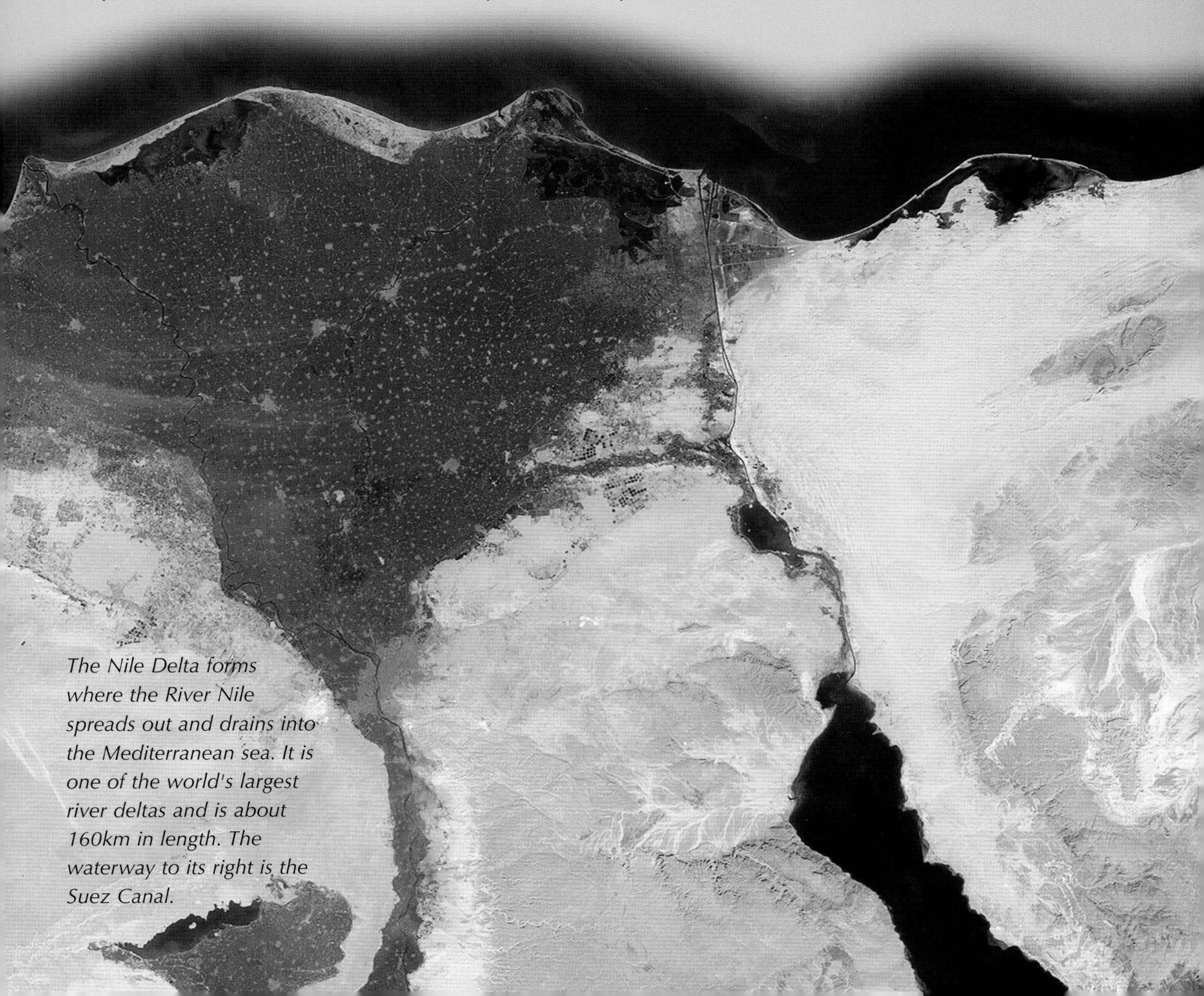

The Nile Delta forms where the River Nile spreads out and drains into the Mediterranean sea. It is one of the world's largest river deltas and is about 160km in length. The waterway to its right is the Suez Canal.

Erosion at the coast

This is a closer view of the arch and cliffs. Can you see the horizontal bedding planes in the cliff?

Waves are constantly hitting the coastline, where the sea meets the land. Over time they can wear away, or erode, solid land structures such as cliffs and rocks. Day to day this can be hardly noticeable, but over time the coastline changes or erodes. Sometimes, storms cause waves of such strength that parts of the cliff will suddenly collapse.

KEY SKILLS

Analysing aerial photographs

Drawing accurate sketches.

IN THE PICTURE

Look at the aerial photograph shown on page 21. The rocks to the front are an arch and the rocks to the back are cliffs. Why do you think the arch has formed? Do you think the cliffs are made of the same type of rock? Using the picture below as a clue, what do you think will happen next to this piece of coastline?

HEADLANDS TO STUMPS

A headland is an area of strong rock that remains after the power of the sea's waves have eroded softer areas of the cliff. Over time, the sea will continue to erode the headland, forming caves in areas of weakness. Eventually, the cave will become so big and deep that the sea will break through to the other side of the headland, through the cave, forming an arch. The waves will continue to erode the arch, while the weather and sea spray will gradually wear away the top of the arch. This wearing away is called weathering. The top of the arch will finally become so weak that it will fall in, leaving a column of rock sticking up at the end of the headland. This is called a stack. In time, it will wear down further into a stump.

Aerial photographs are used to monitor erosion at the coast, as it can have serious consequences.

DRAWING SKETCHES

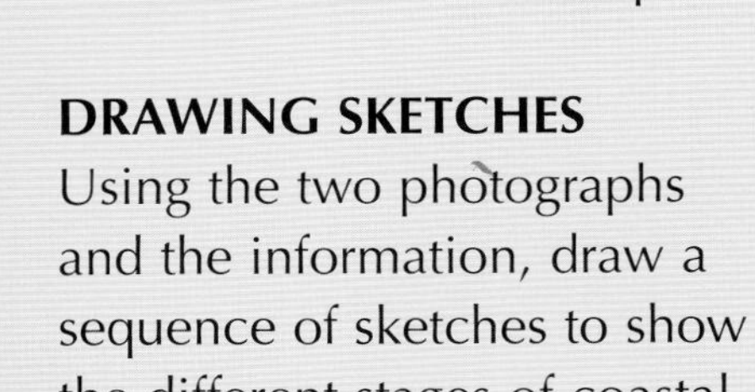

Using the two photographs and the information, draw a sequence of sketches to show the different stages of coastal erosional landforms – from a headland, to an arch, a stack and finally a stump.

WARNING
Never go to the coast without a responsible adult. Always take care around cliffs. Rocks can suddenly crumble and give way.

Depositional features

The sea is constantly eroding rocks and cliffs along the coastline. Caught up in the action of the sea, the rocks and pebbles move along the coastline and gradually wear down into smaller pebbles and sand. This sand needs to go somewhere and much of it is deposited on the beach. However, sometimes it builds up into distinctive coastal features.

This is the end of a spit called Spurn Head in Yorkshire, England. The aerial photograph shows a lighthouse, which is used to warn boats about the location of the spit.

LONGSHORE DRIFT

When waves hit the beach at an angle, they pick up some of the sand on the beach and move it up and along. The sea's backwash brings the sand back to the shoreline ready for the next wave to break and repeat the process. As a result, a lot of the sand drifts along the shore, hence the name longshore drift. Over years, longshore drift forms a long spit of land, stretching out from the shore.

KEY SKILLS

Interpreting aerial photographs

Drawing accurate sketches

Researching a changing coastline

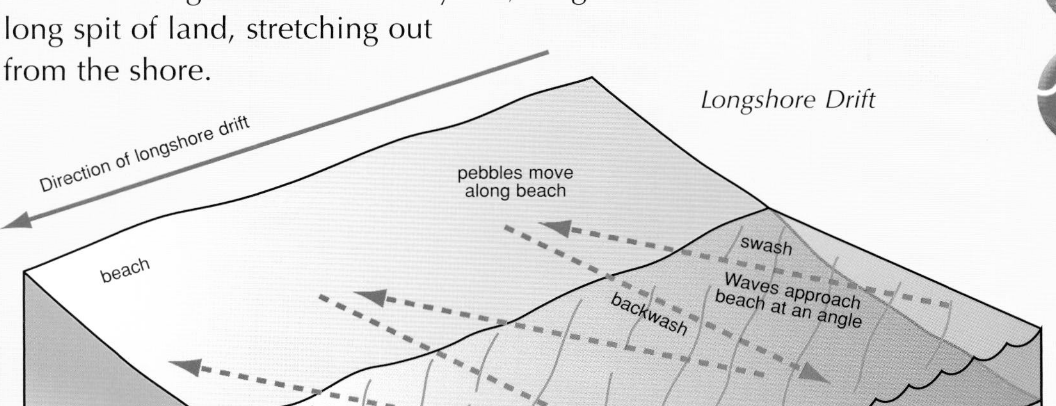

Longshore Drift

SPITS

Look at the aerial photograph on page 22 to understand the features of a spit. Where the spit begins is called the root. On the seaward side, where the waves are breaking, notice that the spit curves in a fairly smooth line, while behind it is more irregular. Look for the widest section of the spit. It should be at the end, proving that the longshore drift is still moving the sand along the beach and depositing it at its furthest point. Look for vegetation on the spit, which will be helping to hold the sand together.

Draw a sketch of the spit and include labels showing the direction of longshore drift. How do you think it looked 30 years ago? Use the Internet to find out why Spurn Head is under threat and to predict how it could look in the future.

SAND DUNES

Sometimes the wind blows sand to the back of the beach, so that it builds up into dunes. You can see dunes in the photograph below. Dunes that are furthest from the sea are the oldest ones. The air is saltly at the coast and it is often windy. Very few plants can grow there, but one kind of plant, called marram grass, does grow well in these conditions. Marram grass has a lattice-work of long roots that help to hold the sand dunes together and stop them being blown away.

Grass has been planted to trap a lot of sand to allow high dunes to develop.

Mountains

Geographers use aerial photographs to examine mountain environments and contrast them with what they know about other environments, such as river deltas or coastal areas.

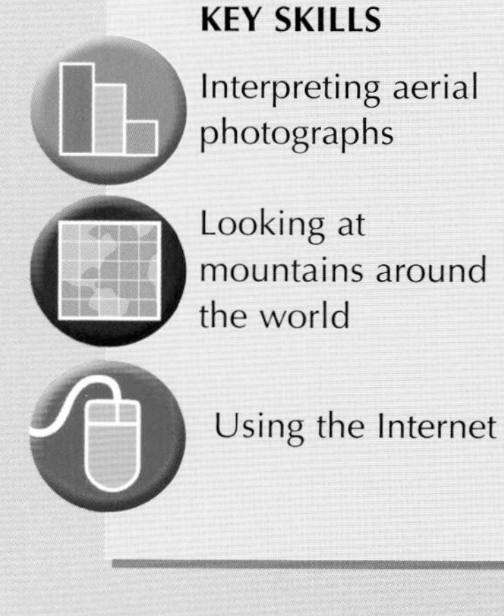

How mountains are formed

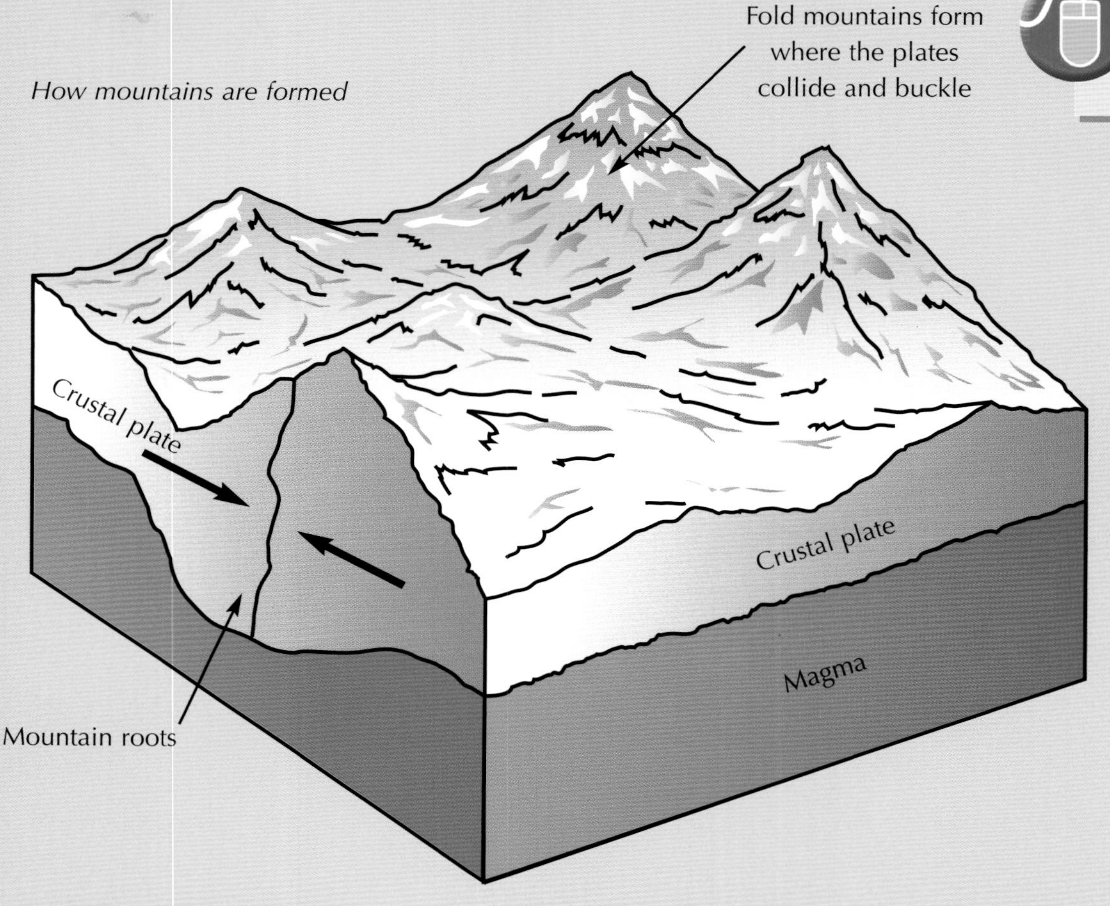

These two crustal plates have moved towards each other. The edges of the plates have crumpled up to form fold mountains.

MOUNTAIN RANGES

Mountains occur all over the world in places where the Earth's crustal plates collide. When this happens, the land is pushed, or folded upwards, creating mountains. Some mountains are continuing to grow higher every year.

MOUNTAIN SLOPES

Look at the photograph on page 25 of the Pyrénées mountains in Spain. What do you notice about the slopes? What do the different layers of rock tell us about the mountain? There are very few plants on the mountainside to hold the soil together. It is washed away by the rain and blown away by the wind. The loose rocks on the side of mountains are called scree. They are rock from the side of the mountain that has been broken down by the harsh weather and fallen down the slope.

MOUNTAIN RANGES

Use the Internet to find a list of the highest mountains in the world. Now, use an atlas to locate these mountains. Plot the mountains on a map of the world. Now, find photographs of the following ranges: Himalayas, Alps and Andes. Look at the sides of the mountains. Can you see evidence of the crusts folding upwards? Can you see different layers of rock? How high up on the mountain are there plants?

HELPING HAND
A blank map of the world is available on www.eduplace.com/ss/maps/

You can see the fold lines in these rocks as they are being pushed upwards by colliding plates.

GLACIERS

As you can see from your earlier investigation, mountainous areas are very high above sea level. This means that they are often cold environments. The snow in these areas creates glaciers, which are rivers of ice. During the colder months more ice forms and a glacier slowly advances down the valley, scouring the rocks beneath it like sandpaper, and moving huge boulders that have stuck to the ice. As the glacier moves, the weight of the ice, and the rocks it moves along, gouge out a wide, steep-sided valley. During the warmer months, some of the glacier may melt.

Search on the Internet to find five glaciers, starting with Glacier Muir in Alaska. Locate them on an atlas and try to find photographs of them. Can you see the glaciers? Find out how they have changed in the last 50 years.

Volcanoes

Volcanoes are created when crustal plates meet. One plate is pushed up to make mountains (see page 24) while the other is forced down into the hot mantle below the Earth's crust. This plate then melts, creating extra mantle which is then forced out through the hole in the Earth's crust to create a volcanic eruption. Some volcanoes erupt every few years, while others erupt only once every few hundred years. Some have not erupted for such a long time that they are called extinct.

This photograph of Mount St Helens, was taken several years after it erupted in 1983. The snow-covered peak of Mount Rainier can be seen in the background.

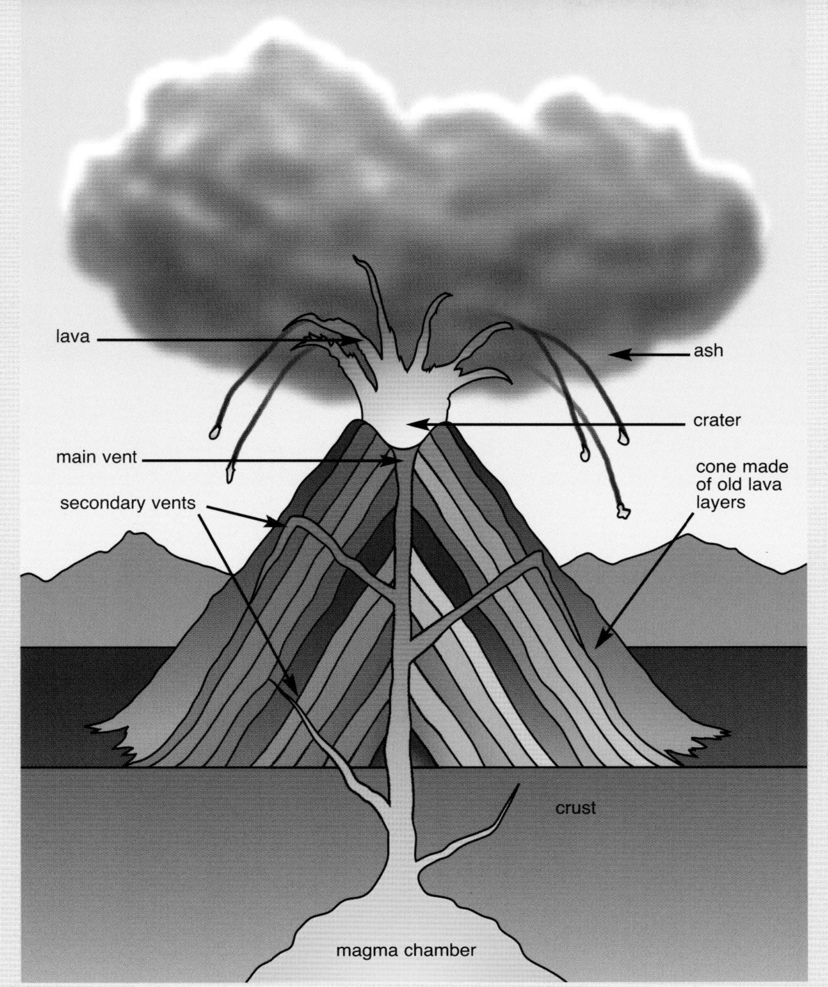

Hot liquid rock pushes up from below the surface of the Earth to the top.

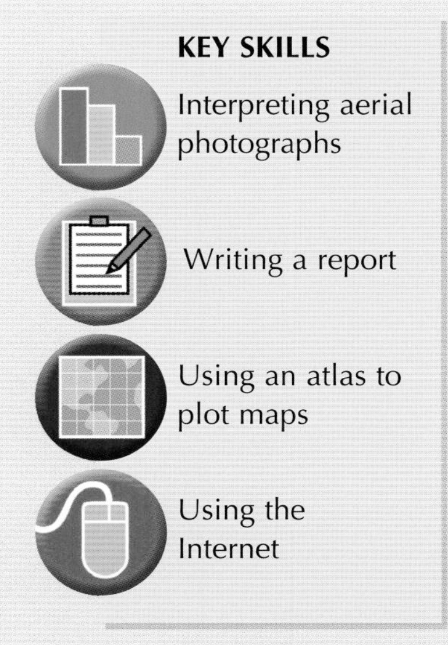

WHEN VOLCANOES ERUPT

Aerial photographs help geographers to monitor volcanic activity. Any changes in the volcano can be seen quickly, and then appropriate action taken. Geographers look for ash or gases and steam coming out from the volcano. If a volcano is about to erupt, warning can be given so that as many people as possible can be evacuated from the area.

MOUNT ST HELENS

Look at the photograph of the USA's Mount St Helens on page 26. This volcano erupted in 1983. Can you see any landscape features that may be a result of the eruption? What does the land around the mountain look like? Find a picture of Mount St Helens taken before 1983 and compare the two pictures. Why do you think the landscape has changed? What impact would these changes have on the environment?

PEOPLE AND VOLCANOES

If people are not given enough warning about eruptions, there can be severe consequences resulting in many deaths. Why do you think people choose to live close to a volcano when it can be dangerous? Carry out some research on two volcanoes that have erupted fairly recently – for example, Mount St Helens and Mount Pinatubo in the Philippines. Collect information and photographs showing how many people were living on the slopes of the volcano at the time of the eruption and how many people died in each case, investigating the reasons for this. You could write up your research as a report.

Now use the Internet and library books to research the locations of five other active volcanoes. You could plot them on the same world map you used on page 25.

Living in valleys

Not all mountainous areas are as inhospitable as the Pyrénées shown on page 24. In some countries such as Scotland and Switzerland, there are so many mountains that people have settled in the ranges. Often, people live in mountain valleys which are the flat regions of land between mountains.

This map shows part of the valley cut by the River Lys in northern Italy.

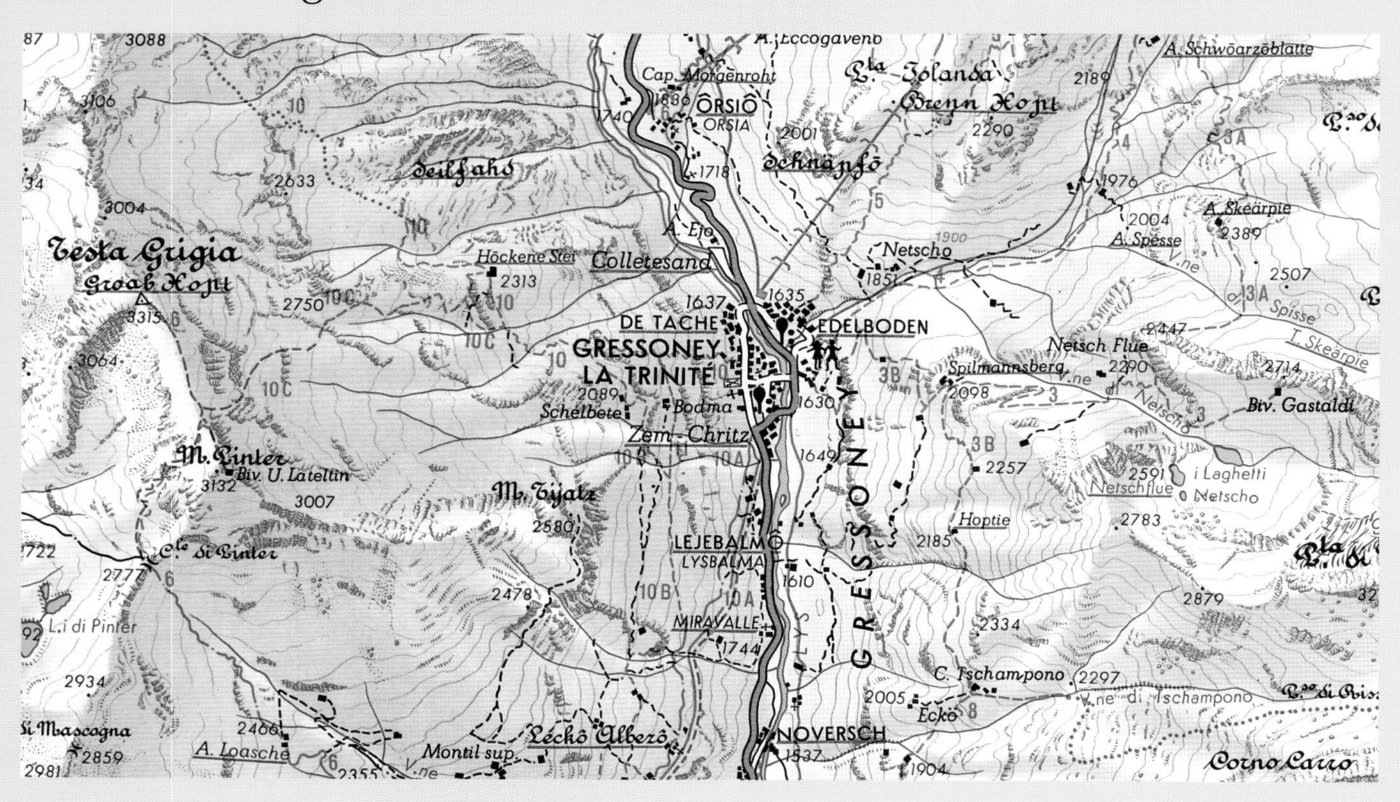

LOOKING AT MAPS

Look at this map of a valley settlement. Find the brown lines. These are called contour lines. They show the map reader how steep and high the land is. If the contour lines are tightly packed together, this means that the slope will be very steep. However, if they are spaced out, then the slope will be gentle or hardly noticeable. Where is the settlement on this map – on the slopes or on the flat land? The flat land is called the valley floor. It is ideal for building on because the land is flat, the soil is fertile and it is warmer here than at the top of the mountain.

VALLEY SETTLEMENTS

Aerial photography gives a vivid picture of how the valley floor is being used. It does not show the contours but you can see the steepness of the slopes of the mountain.

Look at the aerial photograph below. The settlement is identified by the buildings, which stretch along part of the valley floor, following the river. The gentle slopes are more likely to have soil, and some are used for farming. What evidence of farming can you see in the photograph? The higher, steeper slopes are bare rock. Many of the people who live in this mountain settlement earn a living through the tourist industry. What activities could tourists do here?

HELPING HAND
Start your map by looking for mountains, roads, bridges and the river.

Use this aerial photograph to sketch a contour map of the valley settlement. How does it compare to the map on page 28? In which settlement would you rather live?

This is Interlaken, a popular tourist destination in Switzerland. Interlaken is known for its winter sports and adventure sports such as white-water rafting and mountain biking.

Living on mountains

The global population is estimated to be over six billion people. These people all need somewhere to live so if it is possible to live in a place, people usually do. Some areas of the world are so mountainous that people have learned to live on the steep mountain slopes, and make a living there.

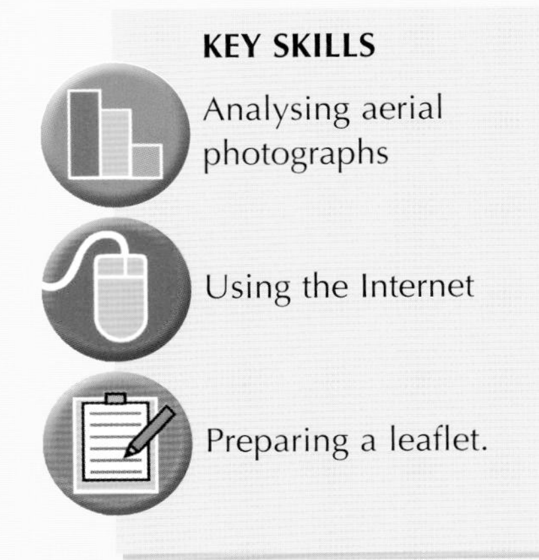

These farmers in Bali, Indonesia, have created steps in the mountainside.

TERRACES

One of the ways that people living in mountainous areas create farmland for crops is to cut steps in the mountainside to make narrow strips of flat land, called terraces. Look at the photograph of terraces in Bali, Indonesia. Every piece of land is being farmed in order to grow the maximum amount of food. Without these terraces, the farmers would not be able to grow enough food to feed their families. Does the photo give you any indication of the climate in Bali? Based on this, what crops do you think the farmers would grow?

These farmers are growing rice. Research other crops that can be grown successfully on terraces – start with tea and coffee. Although tea and coffee will not feed the family, they can be sold and the profit used to buy food. Such crops are called cash crops.

HILLTOP SETTLEMENTS

On page 9, there is a list of the possible reasons why settlements developed. Access to drinking water, building materials, and the ability to grow food and keep animals were all important.
However, in the case of some settlements, defence seems to have been the most important factor. A hilltop site provides just that.

IN THE PHOTOGRAPH

Look at this photograph of Domme in France. Where is the settlement situated? Living on a hilltop may look inconvenient to us now but at the time it would have given the inhabitants of Domme a fantastic view over the surrounding area and allowed them plenty of time to see enemies approaching. Notice how closely together the buildings are, as there was not much space to build on. Now, look carefully at the edge of the village and you will see a wall that has been built around the city. What purpose would this serve?

This aerial photograph of Domme in France shows how the settlement developed on the hillside.

Think about the different mountain areas discussed so far in this book. Why would people want to live in each settlement? Write a leaflet advertising the benefits of living on the slopes of a volcano, in a mountain valley or in a hilltop village.

HELPING HAND
To structure your leaflet, you could do a spider diagram for each settlement. Remember that adverts list only positive points!

Desert environments

People tend to think of deserts as hot, sandy places, but this is not always the case. Some deserts can be cold and rocky. Deserts are defined by their lack of water – most receive less than 250mm of rain per year – plants, animals, soil and population. They are harsh environments to live in but some plants and animals have adapted to survive in these conditions.

KEY SKILLS

Interpreting aerial photography

Doing research

Using an atlas to find other deserts

This is a satellite image of the Gobi desert in China where temperatures can reach a bitterly cold -45°C.

COOL DESERTS

The temperature in some deserts, such as the Gobi in central Asia and the Atacama desert in Chile, can drop to well below 0°C.

Look at this satellite image to understand the conditions there. Can you see any signs of moisture such as a river or a lake? If you look at the bottom left-hand corner of the image, there are green and white patches. These are salt crusts, which formed when water evaporated from ponds. Are there any signs of vegetation? In most deserts there is a lack of soil as this has probably been blown away. It is difficult for new soil to form because of the lack of rainfall. Can you see evidence that people live here?

DESERT RIVER

Look at the aerial photograph above. The flow of the Colorado river has cut a deep canyon into this desert area of the Colorado Plateau in the USA. The surrounding area of the Grand Canyon is dry and stony, and there is no evidence of any other water features such as streams or lakes, which suggests that the area is a desert.

Can you see the Colorado river at the bottom of the canyon? In some places, the river flows at 1.6km below the surface of the desert. This makes it impossible to irrigate the desert. Look closely at the photograph to see some of the different rock layers that the river has cut through. How does the type of rock affect the landscape's appearance? Apart from water erosion, what other form of erosion could help to shape the landscape? Are there any signs of people living in the canyon?

The Grand Canyon is so deep because, as the river cuts down, the land is being pushed up by tectonic forces (see page 24).

DESERT CITIES

Use your atlas to locate the Grand Canyon. Not far from the Grand Canyon is the city of Las Vegas. Find out what technology has enabled people to build a huge city in the desert. Is this a good idea, environmentally? Now use an atlas to find other cities in deserts. How are these cities different from Las Vegas?

Shopping

Most central business districts (CBDs) are full of shops and offices. These businesses have to pay more to be in the part of town with the most expensive land. Land in the CBD is always more costly, so it often leads to the streets becoming tightly packed with buildings and to the construction of high-rise towers.

PEDESTRIANISATION

Cars and lorries still drive through the centre of most towns and cities, but many shopping areas are now closed to traffic. People can walk in these pedestrianised areas without facing dangers from traffic, and the air pollution caused by vehicle exhausts is reduced.

Think about shopping in a pedestrianised area and another shopping area where cars are allowed to drive through. Draw up a chart like the one below. Using the picture below of a non-pedestrianised area as a starting point, compare what you think shopping in these two areas would be like. Your chart should include things such as safety, congestion and transport.

HELPING HAND
Look back to pages 10–11 for more information on the CBD.

	Pedestrianised area	Shopping street
Safety		
Congestion (people)		
Congestion (transport)		
Transport		
Parking facilities		
Affected by weather		

In a busy city centre, shoppers, workers and tourists have to look out for traffic.

The large white building is the shopping centre. Compare the size of the cars to appreciate how big this area is.

SHOPPING MALLS

Things are changing in the world of shopping. The high rents charged to shop owners in the CBD, the overcrowding and congestion, and the difficulty of parking cars and delivery vehicles have led to the development of many out-of-town shopping areas.

The photograph above shows an aerial view of a shopping centre – a complex of shops all built under one roof. What clues show that this centre is built out of town rather than in the CBD? How many car parking spaces can you count? Will it be easier for the shops to receive deliveries from large vehicles in the shopping centre or in the CBD? How will the weather affect shoppers? Will shoppers' safety be better in the shopping centre or in the CBD?

WHICH ONE IS BETTER?

Draw up a table, similar to the one on page 34 to compare the advantages of shopping in the town centre, or in an out-of-town shopping area. Which shops are more accessible for the different groups of people living in the town, like older people, car-owning people, children or office workers? Of all the shopping areas discussed, in which one would you rather shop?

Honeypots

People flock to visit some places perhaps because the landscape is incredibly beautiful, or because there is a building or monument of historical interest there. Areas that attract a lot of people are known as 'honeypots', because visitors cluster round them like bees around a honeypot. Some famous examples of honeypots are the Grand Canyon in the Colorado, USA, Buckingham Palace in the UK, and the Sydney Opera House in Australia.

KEY SKILLS

Interpreting aerial photographs

Analysing information

Observing a local honeypot

Completing a survey; taking aerial photographs

THE IMPACT OF HONEYPOTS

Honeypots or tourist attractions bring visitors to the country who bring money with them. Jobs are created where there may not have been work previously. For example, tourists need to eat in restaurants and cafés or buy food in shops, and often they want to buy souvenirs or even stay in hotels.

However, too many people visiting one place can make it very crowded, and take away from its attraction. Also, the number of people visiting a site can start to destroy it by wearing down the stone it is built from, or the very landscape that is so beautiful in the first place.

THE USE OF AERIAL PHOTOGRAPHS

Geographers use aerial photographs to study the impact of tourists on the area surrounding a honeypot. They monitor the damage caused by too many people walking the same paths day in, day out, and the pollution, such as litter, caused by tourists. They also monitor the impact of cars and buses on the area.

Groups of tourists walk up the steep rock face of Uluru in Australia. So many tourists walk up the rock that it has been badly damaged.

THE COLOSSEUM

Look at the aerial photograph of the Colosseum in Rome below. Can you spot some of the disadvantages of an area becoming a honeypot? How many coaches can you count? Notice the size of the roads leading into the area. Too many vehicles giving off pollution can cause the buildings to become very dirty and can also lead to bigger problems such as acid rain. Notice how there is no grass on this side of the building. This may be because too many people have been tramping along paths, killing the grass, which leads to the soil being washed away.

YOUR LOCAL HONEY POT

Investigate a local honeypot. Find out when it is most busy and how this affects the cost of visiting it in terms of entrance fees, parking fees and rates at nearby hotels. How is the site being managed to cope with the number of people visiting? How is the damage being limited in terms of litter? Try to take or find aerial photographs of the site to substantiate your answers.

Aerial photographs offer a different perspective on tourist attractions. For example, it allows you to see right inside the Colosseum and to appreciate its size.

New houses

Settlements are always changing their shape and size but the people living in them do not always welcome these changes. When new housing developments are planned, many factors have to be taken into account.

This photograph shows the edge of a town. We can tell this by the curved road pattern, houses and fields. The white buildings to the left are modern industry.

EXPANDING DEVELOPMENTS

Look at the vertical aerial photograph on page 38. Imagine that town developers are hoping to expand the town to include the area marked with the yellow X. As you can see, there are already some houses near the X and there is also a developed town on the other side of the main road.

Use the photograph to suggest reasons why this is an ideal location to build new houses. Think about the following:

- How close to the town centre would you be?
- How good would the transport links be?
- What will it be like for the people who are already living in the settlement?
- What about the people in the town?

For the people already living in the town it will mean shops and other facilities, such as doctors' surgeries, being more crowded. There will also be fewer parking spaces and more congestion on the main roads. Does the photograph show any other problems that developers will encounter? What about the motorway? Will the farmers be pleased to sell their land for housing development? How will the environment be affected by the new development? Think about things such as pollution, ecosystems and natural resources.

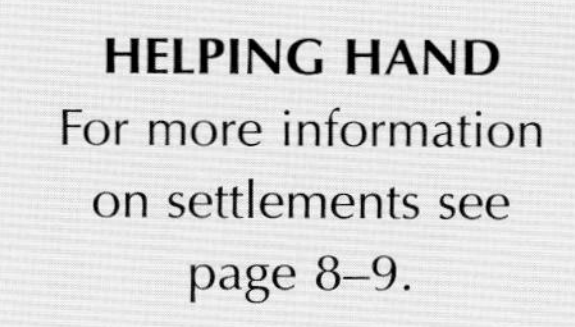

A block of flats in Moscow, Russia. This is high-density living as a lot of people live in a small area.

HOW TO SOLVE THE PROBLEM

With the population expanding year on year, there is a heavy pressure for new homes. Both of these photos show two ways of addressing the same problem. The block of flats will create many more homes using a small area of ground space, while just a few homes on a housing estate would use the same ground space. Can you think of other advantages and disadvantages to living in a block of flats against a house with the same number of rooms?

Limestone quarries

Limestone is a white-coloured rock that formed thousands of years ago from the bodies of millions of sea creatures. People have found many uses for limestone, from toothpaste to cement.

KEY SKILLS

- Interpreting aerial photography
- Analysing information
- Using the Internet
- Preparing a debate

PROPERTIES OF LIMESTONE

Limestone rock is permeable, meaning that it is full of gaps. Water can pass through it so rivers flow underground in limestone areas, sometimes creating amazing underground caves. On the surface, many limestone areas are covered in a distinctive limestone pavement. This is created by rainwater dissolving the limestone in weak areas.

There are few areas of limestone pavement left as it erodes so quickly. This aerial photograph shows you why it is known as 'limestone pavement'.

The quarries are so big that the wheel on this truck transporting limestone is taller than a person.

IDENTIFYING LIMESTONE
Using aerial photography, it is easy to identify areas of limestone by spotting the pavement on the top. The pavement creates stunning scenery, which attracts many tourists. It is equally easy to identify limestone quarries, which leave a huge scar on the landscape. Geographers, environmentalists and planners use aerial photographs, as well as surveys, to study the impact of quarrying in areas of natural beauty.

LIMESTONE QUARRIES
Limestone is the main ingredient of cement, so there is a huge demand for it in the building industry. Limestone is dug out of vast quarries using dynamite to loosen the rocks, and then transported away to factories by large dump trucks.

CONFLICTING OPINIONS
Think about the conflict that is created when an industry wants to quarry in an area of outstanding natural beauty.

Divide a page down the centre. Now write out an argument for both sides; one side in support of quarrying and the other in support of preserving the landscape for future generations.

Who do you think is right? Once you have made up your mind, back up your decision as to whether the quarry should go ahead or not by using case studies from the Internet or photographs of limestone quarries.

Energy needs

The Earth is a very special place and we have a responsibility to look after it. This is called stewardship. We all want warm houses in winter, cool ones in summer, trips out in the car or on buses and new goods, which were made in factories, to bring home. All of these actions will create pollution which causes destruction to our Earth.

This power station is next to a river so that there is a good supply of water.

GLOBAL WARMING

One of the biggest issues affecting the Earth at the moment is global warming. Global warming is the warming up of the Earth's atmosphere over a long time. It is thought that global warming is caused by burning fossil fuels such as coal and oil. These fuels give off harmful gases that trap heat from the Sun and, over time, cause climate change.

Fossil fuels are burned when electricity is made. Look at the oblique photograph of this power station. Note the plumes of smoke that are billowing out of the chimney stacks. Where will the smoke go and what effect will it have on the animals and people that inhale it?

Now look at the oblique photograph of a wind farm on page 43. Wind farms generate electricity without burning fossil fuels so they minimise pollution. What are the other advantages of wind farms?

Can you think of any disadvantages of wind farms? Where is the wind farm situated? Does it look out of place on a scenic landscape? Would you like a wind farm outside your house?

KEY SKILLS

- Interpreting aerial photography
- Analysing information
- Writing a report
- Doing research

PRODUCING ELECTRICITY

Research other ways of producing energy that do not use fossil fuels. Start with hydro-electric plants and solar power. Do not forget about nuclear power. Compare the advantages and disadvantages of each method. Try to look at photographs of each type of energy manufacture.

Wind farms need to be located in windy areas. These are often upland areas as you can see from this photograph.

Glossary

Acid rain
Rain that has a high level of acidity, often caused by pollution.

Alluvium
Sand, silt, mud or other matter deposited by flowing water, such as a river.

Cash crops
Crops sold for money, often on a large scale.

Commute
To travel from the suburbs into the city centre to work.

Contour lines
Lines on a map that joins points at the same height.

Contour map
A map that shows heights above sea level and surface features of the land by means of contour lines.

Crustal plates
Huge, floating plates of rock that form the Earth's crust.

Delta
A mass of alluvium, which is often triangular in shape, found at the mouth of a river.

Deposit
When a river or the sea puts sediment down.

Erosion
The wearing away of the land by water or wind.

Floodplain
The flat land next to a river.

Global warming
A gradual increase in the average temperature of the Earth's atmosphere.

Hydro-electric power
Electricity made from the energy produced by running water.

Informal settlement
Self-built, illegal settlements, locally known as bustees, favelas or shanty towns.

Inner city
Traditionally the area of a city with terraced houses and old industries.

Irrigate
To water the land artificially.

Less economically developed country (LEDC)
A country in which the majority of the population lives in poverty. Most people live in the countryside, but often the cities and towns are growing quickly.

Linear settlement
A type of settlement which has developed in a line, perhaps next to a river or road.

Meanders
Large bends in a river that are formed by erosion and deposition.

Migration
The movement of people, called migrants, from one location to another.

Monsoon
A wind that blows in Asia from the southwest. Monsoons bring heavy rains in the summer.

More economically developed country (MEDC)
A country with much greater wealth per person and more developed industry than a less economically developed country.

Mouth
Where a river enters the sea.

Nuclear power
The electricity made by the energy released during a nuclear reaction.

Nucleated settlement
A type of settlement that has grown around a central point, such as town square or crossing.

Quarry
An open pit from which stone, such as limestone, is obtained by digging, cutting or blasting.

Relief
The shape of the land.

River bed
The bottom of a river's channel.

River channel
The space in which a river flows.

Satellite image
A type of aerial photograph taken by a satellite as it orbits the Earth.

Settlement hierarchy
A way to organise settlements, which shows that there are many small settlements but there are fewer large settlements. Each settlement is vital to the pyramid.

Silt
Small grains or particles of rock, smaller than sand and larger than clay.

Stewardship
People's responsibility to care for the planet to ensure there are enough natural resources for future generations.

Town developers
The people who plan the layout and development of new settlements.

Valley
A long, narrow region of low land between ranges of mountains, often having a river or stream running along the bottom.

Water cycle
The continuous circling of water between the sea, atmosphere and land.

Weblinks

http://earth.google.com
Download free aerial photographs and the corresponding map of any location in the world.

www.webbaviation.co.uk/Portfolio.htm
Hundreds of aerial photographs to look at, just click on the area you want to see.

http://geography.about.com/library/maps/blindex.htm
Go to 'World Atlas & Maps' for a wide variety of maps for every country in the world.

www.niagaraparks.com/webcam/webcam.php
See up-to-date aerial photographs of Niagara Falls.

www.woodlands-junior.kent.sch.uk/Homework/Grivers.html
A really useful site with information and photographs of rivers in the United Kingdom and around the world.

www.bbc.co.uk/schools/gcsebitesize/geography/
Go to 'Geographical Skills', then 'Photos in geography' for all the facts on how photographs are used in geography – you could take the test at the end to see how much you know!

www.fs.fed.us/gpnf/volcanocams/msh/
See the latest aerial photographs of Mount St Helens taken by a webcam.

www.bbc.co.uk/nature/programmes/tv/wildafrica/nile.shtml
Find out all about the Nile river delta, with information and photographs of the longest river in Africa.

Note to parents and teachers:

Every effort has been made by the Publishers to ensure that these websites are suitable for children, that they are of the highest educational value, and that they contain no inappropriate or offensive material. However, because of the nature of the Internet, it is impossible to guarantee that the contents of these sites will not be altered. We strongly advise that Internet access is supervised by a responsible adult.

Index